MOUNTAIN WEATHER

A PRACTICAL GUIDE FOR HILLWALKERS AND CLIMBERS IN THE BRITISH ISLES

David Pedgley

Acknowledgements

This booklet has sprung from lectures given to outdoor activities instructors and courses for Mountain Leadership Certificates, as well as a series of week-long field courses on mountain weather held each summer in North Wales and sponsored by the Royal Meteorological Society. Results from those courses, along with results of other studies described in many technical reports and journals, have been used here to provide illustrations of the wide range of effects that mountains have on the weather. I am grateful to the participants and fellow instructors on those courses for their enthusiastic interest in mountain weather and for helping me to see the kind of booklet that is needed — not only a field guide for all who go into the mountains but also a reference for instructors in mountain outdoor activities. I thank the Director of the National Mountaineering Centre and the Principals of Glenmore Lodge and Benmore Centre for Outdoor Pursuits for their encouragement and help given during the writing of this booklet. For the satellite photographs, I thank the University of Bristol (6G, 22D, 371, 37J, 41A) and the Free University of Berlin (3E, 4B, 7D, 12F, 15B, 15D). I also thank my wife for transforming my rough sketches into the many maps and diagrams. The cloud photographs are my own.

D. E. Pedgley

Contents

Introduction

Our British mountains have plenty of weather. They are almost always windier, colder, cloudier and wetter than low country. What is more, the weather mood can change bewilderingly quickly — from gloomy, lowering skies and driving rain to shafts of sunlight and breath-taking patterns of colour. The weather can delight or endanger life. It can create a mental picture never to be forgotten, or thrust the unwary into a fight for life.

Whether it is to avoid danger or to add to enjoyment of the mountain scene, before starting a day outdoors among the hills it is wise to know what the weather is likely to be. This booklet helps the mountain-goer to get the weather forecast most suited to his needs, and to understand it.

The booklet is in three parts:

Part One describes the kinds of forecasts available and how they can be found out.

Part Two helps you understand the forecast through using weather maps.

Part Three describes and explains some of the ways our mountains get their own weather.

All hill walkers and climbers should read Part One, but if you want to understand the forecast, rather than just know it, Part Two will help you. It is not meant to be more than a start, for an understanding needs a knowledge of why the weather changes. That cannot be learned from a few pages; it comes from experience in using weather maps, and from reading books about the weather. (Some books are listed at the end of this booklet.) The hardest bit is understanding how the mountains make their own weather. Part Three shows you some of the things to look for on a day among the hills. As your experience grows you will become better at modifying the forecast to suit your particular hills and valleys, for no forecast can yet do justice to all the peculiarities of mountain weather.

Know the forecast
Understand it
Be aware of how the hills can change the weather

PART ONE
Weather Forecasts

A weather *forecast* says what the weather is *likely to be* at a given time and place. It differs from a *report* which says what the weather *was* at a given time and place.

There are many kinds of weather forecast. No forecast is right in every detail, but you should strive to get the most reliable one. Use those prepared by a professional forecaster. If you think you can do better, test yourself for a few weeks — that should be enough to show you don't often do a better job. Remember, too, that a forecast can become out of date in only a few hours. If you have any doubt, check the latest forecast.

1. General forecasts

National and regional forecasts, usually for one or two days ahead, and giving general guidance on what the weather is likely to be over the British Isles, are prepared by the Meteorological Office and broadcast on radio and television, printed in newspapers, and can be got from the Post Office's automatic telephone weather service. The forecast regions used in these forecasts are described in a leaflet, *Weather advice to the community,* obtainable from the Meteorological Office, Met O 7a, London Road, Bracknell, Berks RG12 2SZ (phone Bracknell 20242).

Radio and television

Times of broadcasts can be got from the *Radio Times* and *TV Times*. The forecasts for shipping are particularly valuable for the detail they give. Warnings of notably bad weather are broadcast as interruptions to BBC Radio 2 programmes. Snow warnings are on Radio 2 at 4 p.m. Some TV broadcasts have weather maps like those printed in newspapers and used in Parts Two and Three of this booklet.

Newspapers

Most newspapers give the Meteorological Office forecasts, but sometimes only briefly. Some give reports from a list of places and some have weather maps. Most of these maps are forecasts for some time on the day of issue, and they should be carefully distinguished from others that are reports showing the weather at some time on the day before. Remember, newspaper forecasts are not so up to date as those on radio and TV because of the time needed to set up, print and send out the newspapers. Among the most useful are those in the *Guardian,* the *Scotsman* and the *Glasgow Herald.*

2 Forecasts for hill walkers and climbers

Automatic telephone weather service

In many parts of the country, local forecasts prepared by the Meteorological Office are recorded by the Post Office and can be got at the cost of a telephone call. Services available and telephone numbers are listed under 'Telephone information services' at the front of all area telephone directories. Local forecasts for some mountainous parts of Britain are being introduced. Check the directory.

Some forecasts for hill walkers are given on radio and TV. Check for any there may be in your area.

Personal weather service

Forecasts for any given place up to a day ahead can be got by telephone from a number of forecast offices around the British Isles. Check in the area telephone directory for the number to be called. It is often possible in this way to speak to the forecaster himself. If forecasts are not prepared routinely for your mountain area, it may be necessary to order one in advance, to be called for at an agreed time. When ordering, be quite clear about what you want. You will be particularly interested in certain areas and times, and both mountain top and valley weather — winds, temperatures, cloud base and amount, haze, rain, snow and blizzard, and their likely changes with height and time. A fee may be payable for this personal service, otherwise the cost is only that of the call. Remember, the forecaster will not always be able to give all the details you want.

PART TWO
Weather Maps

A weather map is a symbolic picture of the weather over a large area at a given time. Newspaper weather maps and those on TV are much simpler than those used by forecasters, but they show the temperature, wind and weather over, say, the British Isles or Europe. They can be used to help understand the forecast, for they give some idea of *how* the weather changes. Much of our weather is imported; the map shows which way it is coming and how long it will take to reach us.

The next 16 Sections give examples of weather maps for days with a wide range of weathers. Of course, the variations are endless. The usefulness of weather maps grows with experience, so try to understand each forecast in the light of the latest map and its likely changes with time.

Weather maps in this booklet have been drawn like those in newspapers. Each map of Britain shows the weather at 18 places for a particular time. Temperature is given in degrees Centigrade; wind speed in miles an hour is given inside the circles; wind direction is shown by an arrow head flying with the wind; the weather is shown by the following letters.

b	sky largely or wholly cloud free
bc	sky about half cloudy
c	sky largely or wholly cloudy
d	drizzle
f	fog
h	hail
m	mist
p	passing shower
r	rain
s	snow
t	thunder
z	haze

A SCALE FOR RATING WIND STRENGTH (Based on the Beaufort Scale)

Force	Name used in forecasts	Speed range (mph)	General description	Effects on	
				Lake surface	Fresh snow
0	calm	less than 1	smoke rises straight up	like a mirror	none
1	⎫	1-3	smoke drifts, but hardly felt on face	ripples	none
2	⎬ light	4-7	wind easily felt on face; grass and bracken quiver; tree leaves rustle	wavelets but none breaking	none
3	⎭	8-12	heather and small twigs on trees move	some wavelets breaking	a little drift near surface
4	moderate	13-18	small branches move; loose dry grass picked up	some white horses	drifting up to a metre or so
5	fresh	19-24	small trees swaying	many white horses	widespread drifting
6	⎫ strong	25-31	some effort needed to walk; large branches moving; whistling in overhead wires	some spray	some blowing above head height
7	⎭	32-38	walking inconvenient; difficult with pack; whole trees moving	much spray	blowing clouds above head height
8	⎫ gale	39-46	walking difficult twigs breaking from trees	foam in streaks along wind	dense blowing clouds
9	⎭	47-54	crawling difficult; standing and walking impossible, branches breaking from trees		
10	⎫ storm	55-63	progress impossible even by crawling; dragged along by wind; some trees uprooted		
11	⎬	64-72	many trees uprooted		
12	⎭	over 72	great damage		

1. Windy Day

Strong winds are tiresome; they slow you down and make you feel you are fighting a never-ending battle. And it's hard to read a map. Wind strength can be rated on a simple scale showing by how much plants are moved, or waves are built up on a lake, or snow is blown along the ground. The table opposite shows the scale used in some weather forecasts.

Strong winds often last for hours, during which time the air can move hundreds of miles. Hence strong winds often cover large areas. These areas can be seen on weather maps — they are where the *isobars are close together.* An isobar is a line drawn on a map joining places where the weight of the air (atmospheric pressure) is the same. Weather maps show that isobars form patterns somewhat like contours on a topographic map. These patterns change from day to day. On a weather map each isobar has a number showing the atmoshperic pressure corrected to sea level. It is always about 1000 millibars (mbar) — over Britain it is seldom less than 950 mbar or more than 1050 mbar. These differences seem small but they are very important because they tell us the strength and direction of the wind. On map 1A, lowest pressure (about 980 mbar) is over northern Scotland; pressure increases outwards in all directions from this centre. Such a centre of low pressure is called a *depression* or a *low*. The isobars on this day are closest over northern England — hence that is where we find the strongest winds. See also maps 5A, 10A, 14C and 15B for other examples of strongest winds where isobars are closest.

We often want to know not only the wind's strength but also its direction. We say the wind is 'west' when it blows from west to east. Wind direction is that *from* which it is blowing. Winds blow more or less along the isobars such that if you stand with your back to the wind low pressure is on the left and high on the right (diagram 1B). On map 1A there is therefore a south-west wind along the east coast of England and a north-west wind over Ireland. Look at any weather map to test these rules relating isobars to wind direction and speed.

The wind changes from day to day because the isobar patterns change. Lows and other patterns move (often at 20-40 mph or 1000-2000 km a day), grow and fade away over spells of a few days, only to be replaced by others.

If you look at a series of weather maps for Europe and the North Atlantic for a week, say in the newspapers, you will get some idea of how the pressure pattern changes. Map 1C shows a low just west of Ireland, but one day later (map 1D) its centre has moved to just north of Scotland — it has moved about 1000 km in 24 hours at about 20 mph. On map 1C the wind over Ireland was south, but on

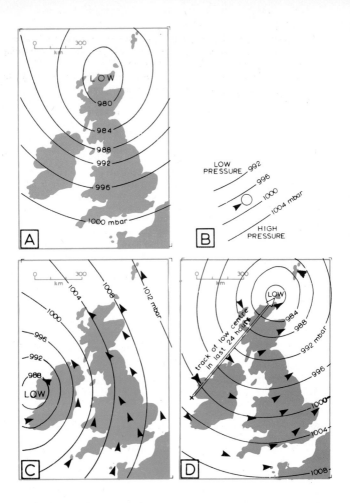

1. A. Pattern of isobars showing a 'low', or depression, centred over northern Scotland
    B. Wind direction related to run of isobars
    C. A 'low' centred just west of Ireland
    D. The same 'low' a day later

map 1D it has changed to north-west. A wind change in this sense — sunwise or clockwise — is a *veer;* in the other way it is a *backing*. Where were the strongest winds on map 1C?

To be forewarned of likely strong winds to come, watch the weather maps for close isobars spreading over you.

2. Calm Day

Light winds are not always a boon. They may be good for sunbathing or fishing, but they don't help you keep cool when climbing.

Weather maps show there are light winds in several kinds of isobar patterns.

1. Map 2A shows an old low centred over England. It started several days ago but now it is filling (its pressure is rising). It is slow-moving and weak. The isobars are well spaced so the winds are light. Old, filling lows often have light winds over large areas.

2. Map 2B shows a centre of high pressure, known as an *anticyclone* or *high.* In its middle the pressure is almost uniform, so there are few isobars and they are far apart. A *high* often has light winds over large areas, and because it tends to be slow-moving the light winds can last for a few days.

3. Map 2C shows two double-centred lows and two highs. There is a *col* over England and Wales — where pressure is almost uniform, and so where there are few isobars and winds are light. The wind is south-west over northern Scotland, but east over southern England.

If you want little wind, watch the weather maps for an old slow-moving low; a high; a col; or anywhere with widely spaced isobars (map 14B is an example).

Even a good breeze by day can fall off at night to near calm, the more so well inland when there is little cloud about. Keep an eye open for widely spaced isobars and a cloudless sky (Section 7).

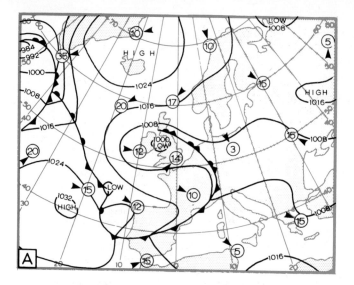

2. A. Weather map for Europe and the nearby Atlantic Ocean, showing a 'low' over Britain

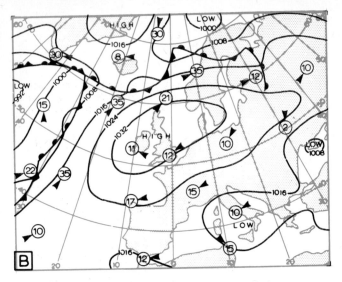

B. Weather map showing a 'high' over Britain

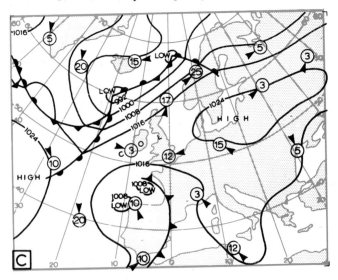

C. Weather map showing a 'col' over Britain

3. Cold Day

We can feel cold not only when the air is cold but when the wind is strong, or we are inactive, or poorly clothed, or wet with rain, or in the shade. In weather forecasts, however, it is the air temperature that is given.

We often think north winds are cold and south winds are warm. This is broadly true, for north winds often come from colder latitudes nearer the pole. On map 3A there is a deep low over the North Sea, and north winds have spread to most of Britain. This map and those of 3A, 6B and 7B show cold winds coming from high latitudes, and map 15B shows cold east winds with a low over the English Channel. Sometimes the wind turns as it comes towards Britain. For example, map 3B shows a pattern of isobars known as a *trough* of low pressure; it lies south-westwards from a low over Scandinavia. If the trough is slow-moving, cold north winds over Scotland turn to west over England. Another example is map 3C, showing a pattern of isobars known as a *ridge* of high pressure; it lies south-eastwards from a high to the north-west of Scotland. North to north-east winds over the North Sea turn to east over England and Wales. Cold north winds can be found a day or two later having turned round to south—maps 3D and 15C are examples.

A weather map gives some idea of where the wind has come from and how quickly it has come. But because pressure patterns change it is often not easy to judge where the air has been in the last few days. The forecaster can give a guide.

Sometimes warm and cold airstreams lie next to each other. The change-over is called a *front*, for it marks the leading edge of one airstream where it replaces another. There are two main kinds of front. Map 3D shows a *warm front;* it is a line along which warm winds (here from the south-west) have reached western Ireland, and will later spread to much of Britain to replace cold winds (here from the south). As a warm front passes overhead the wind veers and warm air starts to reach you. (Photograph 3E is a satellite picture of the clouds at the time of map 3D — see Section 5.) Map 3F shows a *cold front;* it is a line along which cold winds (here north-west) have already spread across Scotland and will replace warm winds (here west) over the rest of Britain. As a cold front passes overhead the wind veers and cold air starts to reach you.

Keep an eye on weather maps for fronts moving towards you. They will give a clue to likely changes to warm or cold winds.

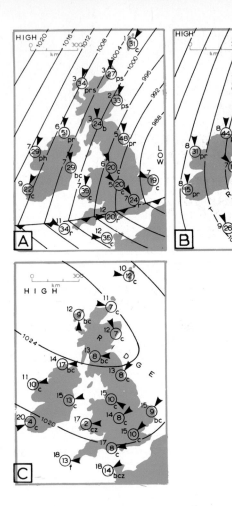

3. A. A deep 'low' over the North Sea, with northerly winds over Britain
    B. A 'trough' of low pressure across Britain, with west winds in the south and north winds in the north
    C. A 'ridge' of high pressure across Britain, with east winds over England, Wales and Ireland

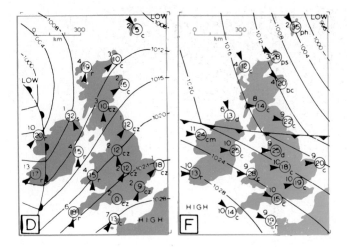

D. A warm front moving east across Ireland, with mild
    south-west winds replacing cold south winds

F. A cold front moving south across Britain, with cold
    north-west winds replacing mild west winds

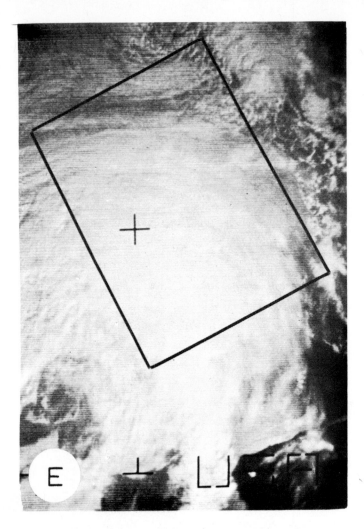

E. Satellite picture for the same time as map 3D (whose outline is shown). The British Isles are completely covered with clouds (see Section 5)

4. Warm Day

We can feel warm not only when the air is warm or moist but when the wind is light, or we are too active, or too heavily clothed, or in the sun.

Warm winds often come to us from the south — map 4A is an example, and photograph 4B is a satellite picture for the same time — see Section 7. (See also map 7C) The weather map may also show warm winds coming from other directions and turning as they approach Britain. For example, map 4C shows a winter high turning to the south-west, with mild north-west winds from the Atlantic covering Britain. The weather need not be cold despite north winds, the more so when clouds are broken and there is some sun to give added warmth.

Keep an eye on the weather map for winds coming from low latitudes. Watch for a warm front passing overhead — it will be followed by warm but often damp air.

Broken clouds make a big difference to the temperature, the more so in summer, because sunshine heats the ground, which in turn heats the air. (Air lets most of the sunshine through; it is not heated much by sunshine.) That's why days with broken or little cloud are warmer than the nights; whereas days with deep, dark clouds are little warmer than the nights, the more so in winter (diagram 4D).

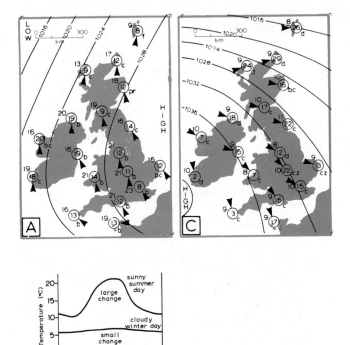

4. A. A high over the North Sea, with warm south winds over Britain

   C. Mild winter north-west winds, with a high to the south-west

   D. Typical temperature changes during the day, contrasting a sunny summer day with a cloudy winter one

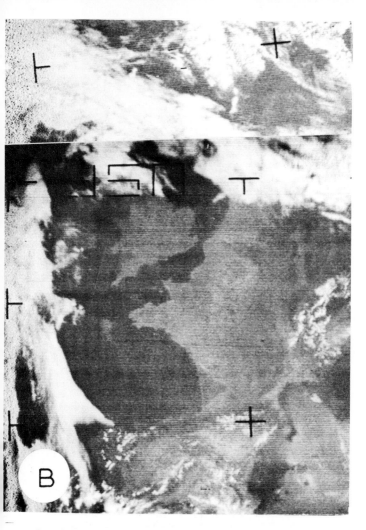

B. Satellite picture for the same time as map 4A. Most of
England, Wales and Ireland are cloudfree, whereas
Scotland is covered. Shower clouds are forming over the
Alps and Pyrenees

5. Dull Day

Dull, sunless weather can be depressing, but for a long, active day outdoors it is often more bearable than endless sun. Dull skies can last for hours or days on end. If, at the same time, winds are strong, this means the cloud sheets are hundreds, if not thousands, of miles across. These vast sheets of cloud show up very well on satellite pictures. Photograph 3E shows Britain wholly covered by cloud sheets at a warm front.

Almost all clouds build up by the lifting of moist air. When air is lifted the weight of the atmosphere on it gets less, so it expands and in so doing it cools (like the cooling of air rushing from the open neck of a balloon). If the air cools enough a cloud of minute water droplets forms, the shape of which shows up the shape of the rising air.

There are three main kinds of cloud that give a dull day.

1. *Stratus* clouds grow in a moist wind blowing over cold land or sea. Winds from low latitudes, or in an area of widespread rain, are very moist and can bring much low stratus cloud (photograph 5A), sometimes with a base below 300m. The air is lifted by countless tumbling eddies caused by the wind blowing over the rough surface of the earth (map 5B and diagram 5C). (See also maps 3E and 5B for moist airstreams full of low stratus clouds; and maps 3D and 14C for examples when the cloud forms in long-lasting rain.) The clouds clear away if the sun heats the ground enough, or if drier air spreads in to replace the moist, cloudy air (usually behind a cold front; see maps 3A and 3E for examples).

2. *Stratocumulus* clouds grow in a moist wind being gently lifted as a whole whilst it streams towards a low or a front (photograph 5D). Winds from the sea in highs and ridges can bring much stratocumulus cloud with a base often between 500 and 2000m (see the south-west winds in map 5E and diagram 5F). The clouds can clear away with a change of airstream, but sometimes a gentle widespread sinking compresses and therefore warms the cloud enough to evaporate the droplets.

3. *Upper clouds* grow within some hundreds of kilometres of fronts and lows when air at heights above about 3000m is gently lifted over wide areas. They lie in vast bands along the fronts (eg photograph 7D), which therefore bring spells of dull weather lasting many hours, or even a few days if they are slow-moving. The clouds clear away as the front moves away, or as it weakens and pressure rises around it. These clouds can be in several sheets. The highest, with bases

between about 6000 and 12000m, are thin and icy *cirrus* (if in tufts or streamers — see photograph 5G), or *cirrostratus* (if in milky sheets). Deeper, denser, greyer ice clouds, with the sun shining dimly as through ground glass, are *altostratus* (photograph 5H), whereas sheets of water cloud like strato-cumulus are *altocumulus* (photograph 5I). Weak fronts and old filling lows may have few upper clouds.

5. A. A drizzly day with *stratus* cloud at about 500m, turning Nant Peris into a gloomy tunnel

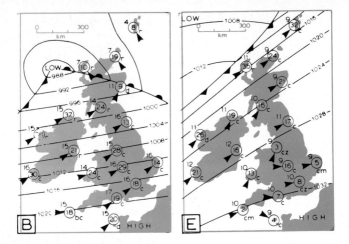

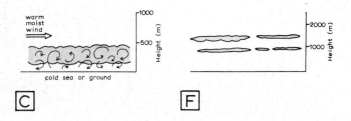

B. Typical weather map when much stratus cloud is brought to Britain on moist south-west winds

C. Tumbling eddies lift moist air to form stratus cloud

E. Typical weather map when much stratocumulus cloud is brought to Britain on moist south-west winds

F. Sheets of stratocumulus cloud form in air gently rising over a wide area

D. Thick, lumpy sheet of *stratocumulus* cloud, with a base at about 1,000m

G. Tufts of fibrous *cirrus* cloud at about 10,000m (with a few small cumulus clouds below at about 1,000m)

H. Multilayered upper clouds ahead of a front approaching western Ireland. The sun shines dimly as through ground glass

I.   Thin sheet of patchy, banded *altocumulus* cloud at about 5,000m over western Ireland – a cloud often associated with fronts

6. Cloudy Day

Cloudy days with only fitful sun are common. Broken clouds drift or rush across the sky, patchily shading the countryside. Such clouds are often *cumulus,* well known for their dark, flat bases and white, domed tops (photograph 6A). They grow when blobs or columns of warm air rise buoyantly from the sun-heated ground or warm sea (diagram 6B). Over land they are mostly a daytime kind of cloud, starting in the morning as the ground warms up and dying away around sunset as the ground cools down again. Over the sea and windward coasts they can be seen both by day and by night when cold winds blow over the warm sea. North winds on the western side of a low are places where they are often found (map 6C, also 3A and 3B). Cloud base is often between 600 and 1200m, higher for drier winds.

Some mornings start cloudless, and then cumulus clouds appear. Each cloud lasts only 5, 10 or 20 minutes, say, but the sky goes on looking much the same as new clouds take the place of old ones. The clouds evaporate by mixing with clear air around them, but sometimes tops will spread sideways into longer-lasting patches that may join to give a sheet over the sky — and so a promising day ends up dull. This is more likely to happen in a ridge or high rather than in a low (diagram 6D and photograph 6E).

The presence of cloud sheets like stratocumulus or altocumulus may not stop cumulus clouds from forming, the more so if the sheets are shallow and the sun is hot. Thus, cumulus and stratocumulus are commonly seen together (diagram 6F).

When small cumulus clouds first start to form they sometimes do so in long rows more or less along the wind. Photograph 6G is a satellite picture of such *cloud streets* over North Wales.

6.   A.   Moderate sized *cumulus* clouds, showing typical dark, flat bases and bright, domed tops. Base about 1,200m and highest tops about 2,500m

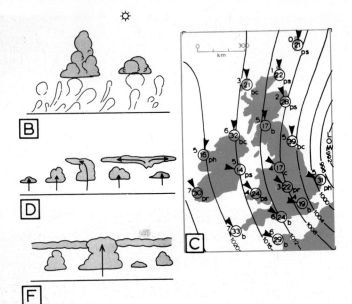

B. Cumulus clouds form by blobs of warm air rising, often from sun-heated ground

C. A day with north winds and widespread cumulus clouds

D. Stages in the spreading of cumulus cloud tops into a stratocumulus layer

F Mixtures of cumulus and stratocumulus clouds are common

E.   Cumulus clouds over Snowdon, with tops spreading into
     shelves of stratocumulus – an indicator that heavy
     showers are unlikely.

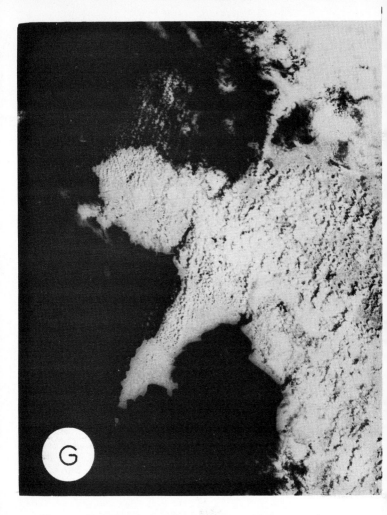

G. Satellite picture of midmorning rows of small cumulus clouds over North Wales. Winds are south-westerly, and clouds that started over Anglesey have been taken out to sea. Apart from these, and some from Lleyn, there are none over the sea because it is colder than the air. A well-marked cloud 'street' lies downwind from the Carneddau and across the Conwy estuary

7. Cloudless Day

Days with little or no cloud are rare. In summer they can give more than 15 hours of unbroken sunshine. Where is such weather to be found? Judging by Sections 5 and 6, it must be away from fronts and lows, and in airstreams too dry for much cumulus cloud to form. The most likely place is a ridge or high where the wind has come a long way over land. Map 7A shows a high centred to the east, with warm, hazy east winds over England and Wales turning to south over Scotland. See also map 4A. A long track from between south and east helps, but even north winds can be cloudless in winter over central England (map 7B), where in summer there would be much cumulus cloud.

When the pressure pattern changes so as to let in a moister wind, not only does the chance of stratus or stratocumulus clouds increase, but so does the chance of cumulus over land by day, the more so in the warmer months. Map 5D, for example, shows the kind of cloudy south-west winds that can spread across when a high that has been giving a cloudless spell moves away. Sometimes the dry wind goes on blowing whilst a front edges its way in; the first clouds to end a spell of cloudless skies are then often patches of cirrus or altocumulus, followed later by dull weather near the front. Map 7C shows a high to the east with warm, largely cloudless south-east winds ahead of a cold front, and clouds creeping slowly east across Ireland and into western Scotland. The long band of frontal cloud is well seen on the satellite picture 7D.

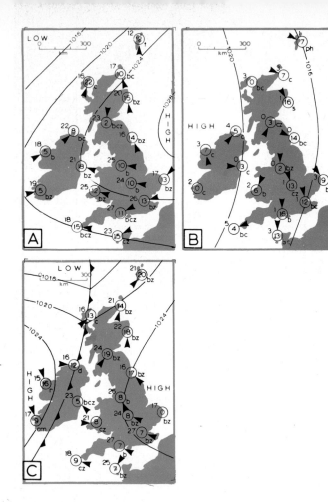

7. A. A ridge of high pressure; sunny but hazy east winds cover England and Wales
   B. Cloudless north winds in winter over central England
   C. Largely clear skies ahead of cloud, rain and drizzle on a slow eastward-moving cold front

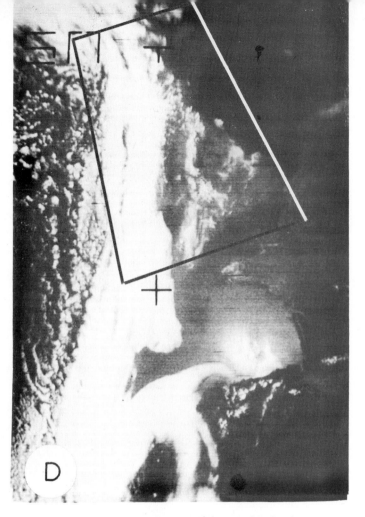

D. Satellite picture for the same day as map 7C, showing
1. Long cloud bands at the cold front
2. patchy cumulus clouds just reaching the north-west corner of the map
3. little cloud over England and eastern Scotland
4. haziness of cloudfree areas compared with 4B
(The Bay of Biscay is bright due to sun glint off the sea.)

8. Hazy Day

Even a day with little or no cloud can be so hazy you are not able to see hills you know are only a few miles away. Those you can see are unusually blue or grey, and much of their detail is lost. The sky may be brownish white, and the sun glows like brass. Such view-spoiling weather can last for days.

Haze is almost always due to smoke. When winds blow gently over large urban or industrial areas, smoke gathers in the air, the more so where its upward spread is stopped at heights of, say, 1000m. The main sources of smoke over Britain are parts of central and northern England, and parts of France and Germany. Hence west winds are often clean, whereas south-east winds can bring widespread haze, the more so when they are in a ridge or high (map 7A is an example).

If hazy air becomes moist, say by flowing across the sea or through a rain area, or it is cooled at night under cloudless skies, the haze can thicken so you can see no more than a mile or two, the more so if there are low stratus clouds as well. The day is then very murky and unpleasant. Haze clears when the wind sets in from a new direction where there are fewer smoky places. Behind a front there can be a quick change, and familiar landmarks can be seen again. On very hazy days it is difficult to tell the kind of cloud in the sky, or its height. In map 7C, clean Atlantic air is slowly edging eastward to replace the haze that extends the length of Britain — from the Channel Islands to Shetland (photogtaph 7D).

Days with gales on the coast can bring another kind of haze — due to countless numbers of salt particles, formed by the evaporation of spray droplets from breaking waves. But salt haze seldom makes it impossible to see less than 10 kilometres.

Watch the weather map for haze coming on light or moderate winds in a ridge or high.

9. Foggy Day

Some mornings you wake up to find yourself shrouded in fog. You may be able to see no further than 50 metres, and trees are dripping wet. This fog may be all the more surprising because the night had been clear, calm and starry. The weather map shows a high, ridge or col with widely spaced isobars; the air is moist but there is little or no cloud. On such a night the ground cools quickly and so does the air, until the moisture starts to condense — some as dew, and perhaps some as fog droplets. By dawn, the fog may be 100m deep or more. As the sun rises in the sky, even though it may not be seen, the foggy air warms slowly and the droplets start to evaporate. By mid-morning in summer, but often later in the weaker winter sun, the fog goes, sometimes with a spell of very low stratus clouds.

In clean air, fog can form in a few tens of minutes, and clear away as quickly. In smoky air, the changes can take hours. Fog can also clear away if the wind picks up, or if a deep sheet of cloud spreads over the top.

Beware a clear, calm night. Be prepared for fog, but don't be surprised if there is only dew.

Wind-borne fog sometimes pours in from the sea. Such fog can come by day or night when warm winds are blowing over cool sea. At sea, the fog clears when the wind freshens enough, or changes direction so that the air becomes drier. Sea fog that has spread inland clears away much like night-time land fog (diagram 9). Weather maps 4A and 7A show fog in Shetland with warm south winds from Europe that have been cooled by the North Sea.

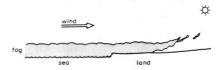

9.   Sea fog spreading inland and dispersing over warm ground

10 Clear Day

Some days are gloriously clear. Far distant hills stand out against blue sky or sharp-edged clouds. The sky looks as though it has been washed — and indeed that's just what's happened. In the past few days the air has passed through the vast cloud sheets of a low, where all the smoke particles became trapped in cloud droplets and fell out in rain. Then the air slowly sank in a ridge or high and reached us after blowing over no smoky places, so it is almost as clean as it can be. Long tracks over the open ocean help to keep the air clean, so in Britain really clear days come with on-shore winds blowing from between south-west and north, often behind a cold front (map 10A). On this map, a cold front has just cleared the south-eastern corner of England, having crossed the whole of Britain from the north-west. Clean west or north-west winds have swept in from the Atlantic.

Clarity of the air is not cut down by clouds in the sky. Indeed, clouds pick out the clarity when they can be so easily seen much nearer the horizon than usual (diagram 10B).

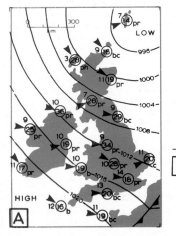

10. A. Clean north-west winds have flooded across Britain
    B. Great clarity of the air lets clouds be seen close to the horizon

11. Rainy Day

Clouds are wet. Walking in cloud, you gather water droplets on your hair and clothes. Plants also gather droplets as cloud blows through them.

Inside clouds, droplets knock into each other and grow slowly. It takes something like an hour to reach the size of a drizzle drop, but not much longer to a rain drop. As a drop grows heavier it falls out. Hence a rainy cloud not only has to last something like an hour or more, it must be deep enough to let drops grow as they fall. Drizzly clouds are usually at least 1000m deep; rainy clouds are much deeper. Hence only deep, long-lasting clouds can give rain or drizzle. Most clouds give no rain.

Map 11A shows a low with a warm front and a cold front. The warmest winds are the south-westerlies between the fronts — in what is called the *warm sector* of the low. In a warm sector there is often much stratus and stratocumulus cloud with outbreaks of drizzle (like photograph 5A). Near the low centre there are often hidden upper clouds with outbreaks of rain. Ahead of the warm front, there is a belt of rain 200km wide or more, moving across country with the front. Near the cold front there is usually a narrower and heavier band of rain. Each front is different; some have little or no rain. If a front is coming towards you, check with the forecast how heavy and how long-lasting the rain is likely to be. Map 5B shows a warm sector, with the low centre moving east towards northern Scotland.

We can see there is a sequence of weather as a warm sector low and its fronts pass overhead (diagram 11B). In the ridge ahead of the low there are likely to be cumulus and stratocumulus clouds by day (like photograph 22D), largely clearing at night. As the wind backs to south before the warm front, upper clouds spread from the west. The first to come are likely to be cirrus and cirrostratus, perhaps with a halo — a white ring around the sun or moon. Then come altostratus and altocumulus, darkening as they deepen to *nimbostratus,* from which rain falls for several hours, with lowering stratus clouds beneath. After the wind veers as the warm front passes overhead, there are the warm, damp south-west winds, and the clouds of the warm sector. As the cold front comes near there can be outbreaks of rain from unseen upper clouds. Rain can stop before or after the wind veers again at the cold front, but as drier air spreads in the cloud base rises, and breaks grow as the cloud changes into cumulus in the north-west winds. The whole sequence may take one to three days, and the details can vary a lot from one warm sector to another. For examples of rain at warm fronts see maps 3D, 5B and 15C; for rain at cold fronts see maps 3A, 5D and 7C.

Signs of likely rain to come are: sheets of cirrus, cirrostratus or altocumulus spreading across the sky, thickening and darkening within a few hours to altostratus. Watch out, too, if formely rainless sheets of stratocumulus darken and lower. Always remember there may be unseen upper clouds that can give rain falling through the lower clouds.

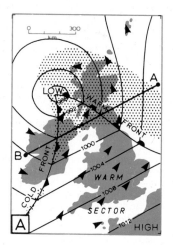

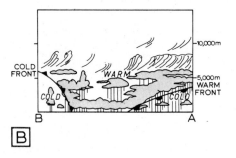

11. A. A 'warm sector' low, with areas of rain shown stippled
    B. Vertical cross-section through a warm sector, along the line AB of map 11A, showing typical layering of clouds and areas of rain

12. Showery Day

Rain sometimes comes in short-lived bursts lasting much less than an hour, with bright or sunny spells between that help us to dry out. Unlike the steadier and more lasting spells of rain that fall from widespread sheets of cloud near fronts and lows, these *showers* come from much smaller clouds called *cumulonimbus*. You can tell these clouds by their tall, swelling, domed and fibrous tops, shining brightly in the sun (diagram 12A). But when skies are more cloudy the tops may not be seen and we must look for the curtains of rain falling from them (diagram 12B). As a cumulonimbus cloud starts to rain out its hard-edged cauliflower-like top becomes streaky and perhaps stretched across the sky in a plume or anvil (diagram 12C and photograph 12D).

Shower clouds grow from cumulus when cold air is heated by warm land or sea. Over land they are commonest in the warm months. If you see cumulus clouds building up quickly during the morning, watch out for showers by midday. They are likely to be heaviest in late afternoon, but die away in the evening. Over sea and windward coasts they are most common in airstreams coming quickly from high latitudes around a large and deep low. Map 12E is an example with a deep low to the north-east of Britain. The corresponding satellite picture (12F) shows that the north-west winds are full of shower clouds. (For other examples, see maps 3A, 3B and 6B. Such showers fall by day and night. Nearer highs, cumulus clouds are usually deep enough to give only light showers (see map 7B).

Always keep an eye on cumulus clouds upwind. Watch for rain curtains cutting the visibility down to a few kilometres; rainbows (if the sun is downwind); dense, streaky spreading tops; cloud caps through which the swelling cumulus tops soon burst (diagram 12G); and sudden, squally winds blowing outwards from dark-based clouds.

After several wettings from passing showers it is helpful to know when they are likely to stop. Look for a change of cloud type from cumulonimbus to shallower cumulus and stratocumulus, say towards sunset or as a ridge of high pressure moves over you.

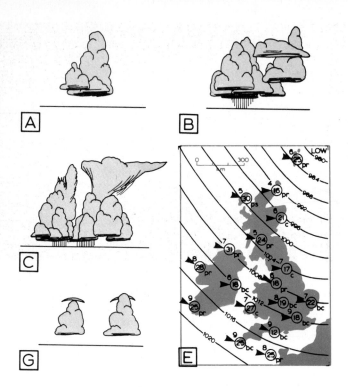

12. A. Massive cumulus tower, beginning to change into a *cumulonimbus*

B. Growing cumulonimbus hidden by cumulus in the foreground, but shown up by its distant dark base and curtain of rain

C. Cumulonimbus clouds with tops becoming streaky (left) and anvil-shaped (right)

E. A deep low to the north-east, with showery north-west winds over Britain

G. Cloud cap, through which the swelling cumulus top soon bursts – a sign of likely showers to come

D. Cumulonimbus, with dark base at about 1,000m, and anvil-shaped streaky top at about 10,000m, accompanied by many smaller cumulus clouds as well as patches of altocumulus

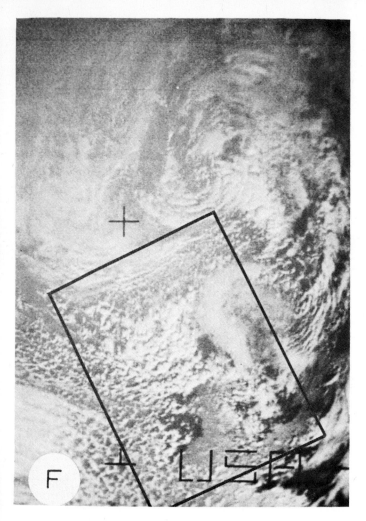

F.   Satellite picture for the same time as map 12E. Apart from south-east England and south-east Ireland, the airstream is crowded with hundreds of shower clouds in an irregular array of blobs, bands and rings

13. Dry Day

Many days give no rain. The clouds are then either too shallow (say, sheets or patches of stratus, stratocumulus or altocumulus — see 5D and 5I) or too short-lived (say, small cumulus — see 6A). If morning cumulus clouds stay flat, or their tops build up to hardly more than twice their base height, showers are unlikely. If cirrus and cirrostratus clouds spread across the sky but only slowly, the more so if they are not followed by altocumulus or altostratus, rain is unlikely perhaps until the next day at the earliest. If stratocumulus clouds stay broken, or they stay above 1000m even if there is a complete cover, there may be a little drizzle.

Highs and ridges often give dry weather (see, for example, maps 3C, 7A and 7C); even fronts within them can give little or no rain because their clouds are too shallow After a dry spell of a few days, watch the weather maps for fronts coming along (maps 7C and 14C). At all times watch for winds changing direction to blow from low latitudes across the Atlantic — they can bring grey, drizzly weather, and so end a dry spell.

14. Thundery Day

Thunder is rare in Britain; most places have only 5 to 10 days a year. At worst it is only noisy, but lightning can be a danger. Thunder is the noise made by giant electric sparks we call lightning. Because sound travels much slower than light, the time between flash and clap is greater the further away is the flash — three seconds for every kilometre. So, a 12-second time gap means the flash is four kilometres away, and if the thunder cloud is coming towards you on a 30 kph wind it will be overhead in 4/30 of an hour (8 minutes). Thunder can seldom be heard more than 20 km from a storm, but lightning may be seen more than 100 km.

Lightning tends to strike upstanding things like trees and tall rocks. If they are wet with rain, the electric current is likely to pass through the water; but if they are dry on the outside, lightning can pass through moisture in the rock cracks, or through tree sap, sometimes explosively because the moisture boils at once, splitting open the rock or tree.

On a ridge or summit in thundery weather you may hear crackling noises, and feel your hair stand on end as though it were being pulled. It may even glow with 'St. Elmo's Fire.' Retreat then to a less exposed place where the risk of a nearby lightning strike is less.

Thunderstorms usually grow out of large and energetic showers (diagram 14A). They often come with hail. If you see cumulonimbus clouds growing large, the more so if they are giving heavy rain and hail, watch out for lightning and thunder. A cloud often gives thunder for less than half an hour, but as new showers grow and pass by on the wind, a storm can last for hours at any one place.

Over land, thunder is most likely in the afternoon, on days with plenty of showers (section 12). More widespread thunder, with outbreaks of heavy rain, can come by day or night in summer when a cold front from the west has become slow-moving after a warm spell (map 14B), or when a low or trough is slow-moving over or near Britain (map 14C). Watch out for a cloud called *castellanus;* it looks like rows of turrets sprouting from a common base (diagram 14 D). If these clouds spread and grow along with other sheets as the front, low or trough gets near, there is a fair chance of thunder to come.

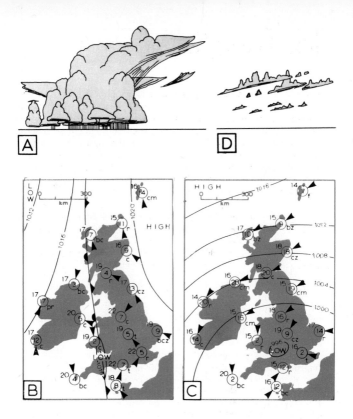

14. A. Massive group of showers, likely to give thunder and hail
    B. Slow-moving summer cold front with outbreaks of thundery rain
    C. Slow-moving summer low over southern England with widespread thundery rain
    D. *Castellanus* cloud – a sign of likely thunder to come

15. Snowy Day

Falling snow makes a pretty scene, but life outdoors becomes rough, progress is slowed as dangerous hollows get blanketed, and things may be hard to see even only a kilometre or two away.

The kinds of weather patterns that lead to snow are much the same as for rain; it's simply that the air is cold enough to let snow reach the ground before it can melt. Long spells of snow come with lows and fronts, and snow showers with cold airstreams. Winds can come from any direction, but those from between south and west seldom bring snow for more than a few hours. A north wind with a large, deep low over the North Sea can bring heavy snow showers to north-facing coasts and hills (such as in maps 3A, 3B, 3E, 6B, 12D and 15B), but cold east winds can give snow showers even with high pressure (often centred over Scandinavia — see map 15A). A low moving east over England can bring long-lasting snow and east winds to Scotland, but if the low crosses northern France, England and Wales have the falls. Map 15C is an example and the associated spiral masses of cloud are well shown in the satellite picture 15D.

Often only a small change in temperature will mean a difference between snow and rain. With south-east winds ahead of a slow-moving front, snow can go on falling for more than a day (map 15E), slowly changing to sleet and then rain as winds turn to south-west behind the front.

If you look at a snow flake you will see it is a loose jumble of fragile ice crystals, many like needles or six-pointed stars. These crystals grow inside the clouds and tangle together as they fall. In very cold weather, the flakes are small, but near melting point they are sticky and sometimes join into large clumps.

Snow can fall when the air is warmer than melting point. This happens when the air is dry, for then the flakes slowly evaporate as they fall and the resulting cooling keeps them below melting point. Snow showers in very dry north winds can be seen at temperatures as high as $7°C$.

If rain is forecast, but the air is cold and dry, snow may fall instead.

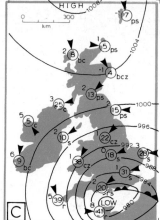

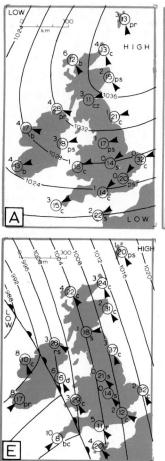

15. A. Winter east winds with snow showers

    C. Winter low with widespread snow over England and Wales

    E. Slow-moving winter front bringing long-lasting snow, turning to sleet, rain and drizzle near the front

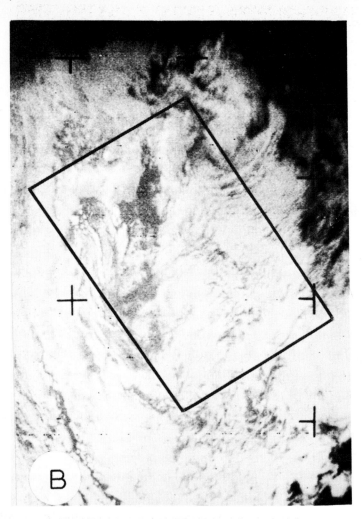

B. Satellite picture for the same time as map 15A. The snow shower clouds have a tendency to be in lines along the wind. Snow-covered ground looks much like clouds, but an area of clear skies (and highest temperatures) covers north-west Scotland

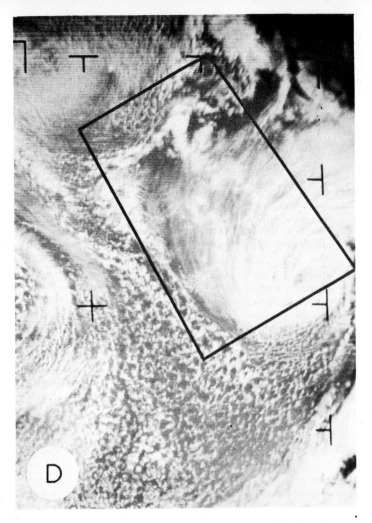

D. Satellite picture for the same time as map 15C. Most of England and Wales are covered by mutilayered clouds. Elsewhere there are shower clouds like those in 12F, but Scottish skies are largely clear, showing the ground is now covered – even Skye

16. Blizzard

Snow and strong winds can lead to a bad day. The wind blows lying snow into swirling clouds — a *blizzard*. Going forward becomes very hard in the choking, blinding snow. When freshly fallen, snow is loose and easily blown in a cold dry wind little stronger than Force 4. Older snow, crusted after lying a few days or partly thawed and refrozen, drifts less easily. Because there are calmer patches behind upstanding things like plant tufts and rocks, blowing snow tends to settle there, and form drifts stretching down wind (diagram 16A). In time, these drifts grow until the wind blows smoothly past them. Just in the lee of a ridge, swirling eddies keep the ground more or less free of lying snow for a while, but even there a long-lasting blizzard can lead to a smooth drift a metre or more deep. Although such a place might be taken as a shelter from the wind, you run the risk of being buried later. Alongside upstanding rocks, strong winds sweep the ground free of snow (diagram 16B), whereas even in front of them there can be a place where the wind weakens enough to let a drift gather (diagram 16C).

Be on the watch for a blizzard when both snow and strong winds are likely; maps 15A and 15B are examples with drifting even over low ground. A bad day can be expected when snow falls on the northern side of a winter low moving slowly east. But even one heavy shower in a strong, cold, dry airstream can turn a fine and bracing day into a nasty one.

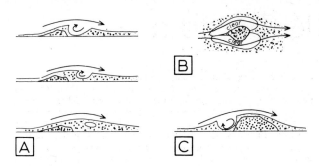

16. A. Stages in the growth of a snow drift behind a boulder
    B. Plan view of a boulder with wind-swept scoops on either side
    C. Drift forming upwind of a boulder

PART THREE
Weather on the tops and in valleys

In the remaining sections we look at those days when weather in the mountains differs from that over low country. This often happens, so when listening to or reading a forecast for the general public it is necessary to make some allowance for the differences. Sections 17 to 31 give some guidance on how this can be done for mountain tops, and Sections 32 to 46 for valleys.

17. Windy Top

Mountain tops are nearly always windier than open, low country. There are two reasons for this. Firstly, even over open country the wind very often strengthens with height. Secondly, the wind strengthens as it blows around and across mountains.

The strengthening with height can be seen easily in the shapes of shallow cumulus clouds first forming in the morning: they often tumble forwards as the wind pushes the upper parts of masses of rising warm air faster than the lower parts (diagram 17A). Tall cumulus clouds can become leaning towers (diagram 17B). Because there is this strengthening with height, the wind speed on an isolated 1000m *peak* is about twice the speed over open, low country.

The wind blows over rather than around a mountain *ridge*. On many days it does this in a way like water flowing over a boulder in a river. Just like the water speeding up as it becomes shallower over the boulder, so the wind strengthens, the more so when the lowest clouds are stratus or stratocumulus and not cumulus (diagram 17C). On such days the wind over a ridge can be three times the speed over open, low country.

Look for these strong ridge-top winds on days with much low cloud, as in the warm sector of a low (maps 4B, 5B and 11A); or when a warm front is coming towards you and is less than about 100 km away (maps 3D and 15C); or around a strong high (map 15A). The strongest winds in a warm sector tend to be in a long belt about 100 km wide just ahead of the cold front (map 27A). When the winds in this belt are forecast to be strong or gale force over low ground, they can be hurricane force and very dangerous over high ridges. On the other hand, on showery days with large cumulus and cumulonimbus clouds, the wind blows more easily over ridges. The strengthening with height is then less, but still needs to be heeded, the more so when winds are already strong at sea level (as in maps 3A, 3B, 6B and 12D).

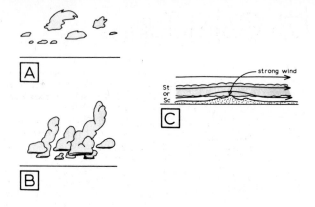

17. A. Forward-tumbling small cumulus clouds when the wind strengthens with height

    B. Leaning cumulus towers

    C. Strengthening of winds over mountains because lifting is less at greater heights

18. Calm Top

Mountain tops are not always windier than open, low country There are two reasons for this. Firstly, the wind can sometimes weaken with height even over open country. Secondly, an eddy with weak winds can form in the lee of a ridge.

A weakening of wind with height is most likely above about 500m, but it happens on only a few days, for example in east winds on the south side of a high moving eastwards. A forecaster will be able to advise on whether such a weakening is likely.

There is often a leeside eddy when the wind blows across a crested ridge (diagram 18A). The wind blows up the windward slope but leaves the ground at the crest. On the leeward slope the wind blows in the opposite direction, often weakly and fitfully, and the join between the two winds can be very sharp. Much the same thing can happen at the windward edge of a flat-topped ridge (diagram 18B), or even on the windward slope, where the steepness changes suddenly (diagram 18C). When snow is blowing along the ground, these quieter places can be sought out for making a safe shelter, at the same time remembering that deep drifts can build up there. Indeed, on the leeward side of a ridge the drift becomes an overhanging cornice (diagram 18D).

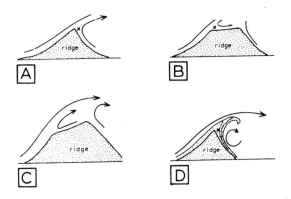

18. A. Lee-side eddy, with light winds at 'X', just in the lee of the ridge crest
     B. Eddy at upwind edge of flat-topped ridge
     C. Eddy in the lee of a change in slope
     D. Snow cornice overhanging leeward side of a ridge

19. Cold Top

Even though a forecast is for warm weather, it must not be forgotten that it nearly always becomes colder the higher we go. The lingering of high-level snow is a reminder of this cooling with height. The rate of fall of air temperature with height is known as the *lapse rate* — on average it is about 1°C for every 200m. Using this lapse rate and the valley bottom temperature we can estimate the temperature at any height above the valley. For example, suppose the temperature in a valley at 300m above sea level is 10°C; what is the temperature on a summit at 900m? The height difference is 600m, so the temperature difference is 3°C; and the summit temperature is therefore 10 − 3, or 7°C (diagram 19A).

On some days the lapse rate can be as large as 1°C for every 100m. Windy days and cumulus clouds are likely to have such a large lapse rate, for the air is then well stirred. Take an example on such a day. Suppose the temperature is 7°C in a valley at 200m above sea level on a sunny but windy winter afternoon. At 1200m the temperature would be about 7 − 10, or −3°C (diagram 19B). Thus, the summit would be below freezing and, bearing in mind the likely strengthening of wind with height, the summit weather would be severe, despite the pleasantness of the valley, the more so if showers are blowing along the wind.

On this day, 0°C would be reached at about 900m. This height is known as the *freezing level;* it can vary in height greatly from day to day, and is on average lower in winter than in summer, and in northern Britain rather than the south. The forecaster will be able to advise on what the height of the freezing level is likely to be.

Inside clouds, the lapse rate is seldom more than about 1°C for every 200m.

When the wind is forced to blow across a ridge, summit temperatures can be lower than in air at the same height over low country. This is most likely where a sheet of stratus or stratocumulus cloud has been deepened over the ridge Although the cooling is usually small it can lead to unexpectedly icy weather when temperatures would otherwise be just above freezing.

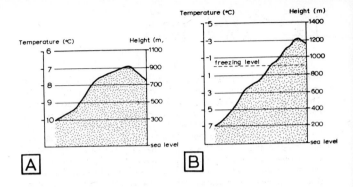

19. A. *Average* temperature lapse rate – 1°C fall for every 200m gain in height

  B. *Large* temperature lapse rate – 1°C for every 100m gain in height

20. Warm Top

Sometimes a mountain top is surprisingly warm judged by the forecast for low ground. There are three main ways in which this can happen.

1. After a windless night with little cloud, instead of the air being cooler at greater heights it is warmer. We say there is a *temperature inversion*. Quite often the temperature can *increase* by 5°C in going up only 100m, and a more usual lapse rate may not start until a height of 200 or 300m above the valley is reached (diagram 20A). The effects of a *ground inversion* can sometimes be seen in frosty weather when white hoar frost lies only on the valley bottom and not the higher slopes. Estimating summit temperatures is not easy when there is a ground inversion because you will not know how strong or how deep it is. A forecaster will be able to advise. Suppose there is an inversion of 5°C over a depth of 300m, and we use the first example in Section 19 — 10°C in a valley at 300m above sea level; what is the temperature at 900m? Given this inversion, the temperature at 600m is 10 + 5 = 15°C. Then taking a lapse rate of 1°C for every 200m rise, the temperature at 900m is 15 − 3 = 12°C. Contrast this with 7°C for the earlier example.

2. On some days, even windy or cloudy ones, there can be an *inversion aloft*. Diagram 20B shows an example where there is an inversion of 3°C between 600 and 1000m. Both above and below this inversion layer the lapse rate is about average. Unlike a ground inversion, which usually forms on a clear, calm night and goes away as the morning gets warmer, an inversion aloft can occur by day or night, and it is not easy to tell if one is there. One clue is a sheet of cloud or haze through which mountain tops poke like islands. The top of a cloud or haze sheet often lies at the bottom of the inversion. If the inversion rises or sinks past a mountain top there can be dramatic changes of temperature there. The two main ways whereby an inversion aloft forms are:

   (a) air below is cooled — say when it comes from low latitudes or from hot land across a cool sea, as in the south-west winds of a warm sector low or the east winds of a summer high over Scandinavia (map 7A);

   (b) air above is warmed — as by slowly sinking in a ridge or high.

3. On sun-facing slopes when there is little wind, the ground becomes warm, the more so where it is rocky, and warm gusts rise to the tops (diagram 20C).

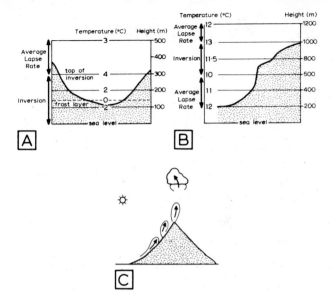

20. A. A valley-bottom temperature inversion
    B. A temperature inversion aloft
    C. Sun-heated slope with little wind – gusts of warm air rise past the summit

21. Dull Top

Many clouds over Britain have bases below 1200m, so it is no wonder our mountain tops are often in cloud, the more so in winter. But days when the forecast speaks of broken or even little cloud can be dull and sunless over the hills for hours on end. The reason is that mountains can make their own clouds. As we saw in Sections 5 and 6, most clouds are made by the lifting of moist air — in the tumbling eddies of strong winds, or the buoyant masses of warm air on sunny days, or the widespread gentle lifting near lows and fronts. But even if there are no clouds forming by these kinds of lifting, there are others that form when air rises over a mountain barrier.

Air rising up the windward side of a ridge can give cloud if it is moist enough. You can watch clouds forming there in more or less the same place for hours on end. They start as small shreds, rise, grow, and join to form a mass that seems to sit on the highest ground. On the leeward side, where the wind is sinking, the cloud breaks up again because it is being compressed and warmed as it sinks (diagram 21A). (An example of air being warmed as it is compressed is the pumping up of a cycle tyre.) The cloud may be just a small wave-like cap on an isolated hill (diagram 21B), or a vast grey shroud of stratus or stratocumulus over a broad highland, beneath which the deep valleys run like gloomy tunnels (diagram 21C and photograph 5A).

Watch out for dull tops whenever there is a moist wind blowing over the mountains — the more so when the weather map shows the wind has come across the sea from low latitudes (map 5B), or there has been rain for several hours (maps 3D, 14C and 27A). Expect the cloud base to rise, even to above the tops, if the wind becomes drier, as it may well do if a fitful sun makes the day warmer, or if the wind direction changes to blow from higher latitudes (say, after a cold front has passed — maps 3E and 15C), or from a warmer country in summer.

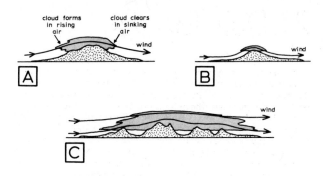

21. A. Cloud continously forms and clears from the same place as moist air crosses high ground

    B. Wave-like cloud cap on an isolated hill

    C. Vast cloud mass over highlands, with deep valleys turned into tunnels

22. Cloudy Top

Apart from wholly sunless days, the tops can be cloudy with only fitful sun even when the forecast is for 'bright intervals'. On a day when cumulus clouds are widespread over low ground, those carried on to windward slopes can grow more strongly in the air forced to rise over the high ground, or they may even be set off there, Either way, cumulus clouds, and any stratocumulus formed by the spreading of their tops, tend to be deeper and longer-lived over mountains than over low ground, with the mountain tops passing in and out of cloud (diagram 22A). Even on days with little or no wind, cumulus clouds tend to form first and grow larger over mountains compared with low ground. There are several reasons for this. After a clear, quiet night there may be a ground inversion through which the mountain tops poke. Cumulus clouds cannot start over the low ground until the inversion has been destroyed by the sun's heating, but over the higher ground they are not thwarted in that way (diagram 22B). Moreover, sun-facing slopes warm up faster than flat ground, so masses of warm air can break away from there and rise to give cumulus clouds (diagram 22C). Photograph 22D shows a satellite picture of cumulus clouds over the mountains of Skye but not over the sea. Remember, the base of cumulus clouds usually goes up during the daytime, but it may well come down again in the late afternoon on windward coasts, or indeed at any time of the day if the wind becomes moister.

A *banner cloud* (diagram 22E) can form in the lee of an isolated mountain around which the wind blows freely but where a leeside eddy raises air further than in any eddies there may be moving on the wind. Cloud forms at the top of the eddy where it turns downwind, its edges being ragged because it mixes with cloudless air around it.

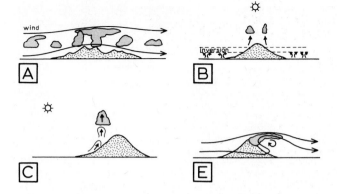

22. A. Cumulus clouds tend to be deeper and longer-lived over mountains than over low ground

   B. Ground-heated air does not rise far enough beneath the inversion to form clouds, but it does over the higher slopes

   C. Cumulus clouds forming over sun-heated slopes

   E. Banner cloud in the lee of an isolated mountain

D. Satellite picture of midmorning cumulus clouds over the mountains of Skye, illustrating the tendency for such clouds to first appear over mountains after the sun starts to heat the ground. Elsewhere there are patches of stratocumulus and altocumulus clouds – best seen over the sea, but mixed with cumulus over the mainland of Scotland. Winds are north-westerly

23. Cloudless Top

On some days with a general forecast of dull, sunless weather, you can be on the tops in hours of bright sunshine. There is hardly a cloud overhead, yet a 'sea of clouds' below. The highest peaks and ridges come through the soft, quilt-like sheet as though they were islands, and the great clarity of the air lets you see far into the distance (diagram 23A).

This kind of weather is not common but it can happen when a large high sits over Britain and all the clouds lie beneath an inversion aloft that keeps below the mountain tops. Winter is perhaps the most likely time of year to find this weather because the heat of the summer sun tends to keep the inversion too high, except perhaps near windward coasts. At sun-facing slopes on windless days, the edges of the cloud sheet can be seen breaking up, and shreds are carried upwards in air rising from the warm ground (diagram 23B). If you stand with your back to the sun and look down into a cloud-filled hollow you may see a *Brocken Spectre*— your own shadow stretched out over the cloud top, and around the shadow of your head (but not of your companions') are coloured rings known as a *glory*.

Seen from a valley, the grey sky does not look promising, but if it is not too dark, the weather map shows a high, and there is no drizzle despite a low cloud base, suspect that the cloud is shallow and the mountain tops clear, the more so if there is a gap through which no higher clouds can be seen. The discomfort of climbing through cloud may be rewarded by the bright, sunny tops. Even if the tops are still just in cloud they can clear later if the inversion sinks a little. On the other hand, watch the cloud top below you to see it does not rise and swallow you up unawares. The forecaster will be able to advise on this kind of cloud but, remember, its top can be domed up by 300m or more compared with that over low ground.

Sometimes a sheet of very low cloud does not reach the top of a ridge because the wind is blocked (diagram 23C). The leeward side is then free of this cloud (although there may be others at greater heights).

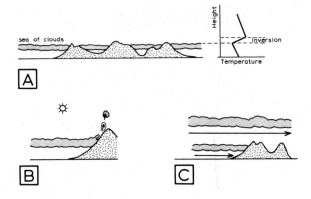

23. A. Sea of clouds, with mountain tops like islands above the temperature inversion

   B. Shreds rising from the edge of a sea of clouds where it lies against a sun-heated slope

   C. Lower cloud sheet blocked by mountains, giving a ·cloudfree lee side

24. Hazy Top

Days with widespread haze may, or may not, bring hazy weather to the tops. If the haze is shallow, as it is more likely to be in winter than in summer, the highest ground may be above the haze top (diagram 24A). But if the haze is deep, even the highest tops can be enveloped. In summer, a shallow haze at dawn can deepen to cover the tops by mid-afternoon (diagram 24B). A change in wind direction can bring in a deeper haze from far away.

Haze is thickest just below cloud base. When climbing on a hazy day, the view can become poorer as you get closer than about 300m below cloud base (diagram 24C). This is because the minute smoke particles that make up the haze grow into small droplets by attracting water from the air when it is moist enough — as it rises towards cloud base. On very hazy days the base becomes so fuzzy it is not easy to tell when you are in or out of cloud.

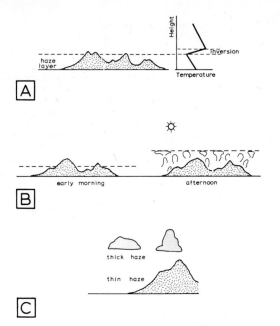

24 A. Haze layer trapped below a temperature inversion

B. Haze layer deepening during the day as rising blobs of warm air mix haze into clear air above the inversion

C. Haze is thicker in the moister air near cloud base

25. Foggy Top

Clouds enshrouding the tops are often called 'mist'. In them you can see about 100m, but it is rare not to be able to see as far as 50m. This is also called *hill fog*. Because clouds down on the hills are common in Britain, and they are a hazard to the safe crossing of high ground, a constant watch should be kept for tops getting covered.

1. If there already are sheets of cloud not far above the highest tops and the wind is changing to become moister, expect the base to come down. Keep an eye on the highest tops, for they are likely to be the first to get covered. Often there is plenty of warning, for the base may fall only a hundred metres an hour.

2. If there is a spell of rain, the base may fall more quickly. After a few hours of steady rain, windward slopes may stay covered to below 300m above sea level, until the passage of a front, say, brings in a drier wind.

3. On a showery day, there are patches of moist air passing by on the wind. When one of these patches flows over high ground, low clouds form quickly but they may last much less than half an hour. With showers about, expect to see not only the highest peaks coming and going, as the clouds drift by, but also some of the lower hills (diagram 25A).

4. On a cumulus day, when the sky starts blue in the early morning, the first clouds are often the lowest, so don't be suprised to see the tops soon cloud over. The forecaster will be able to advise on the likely base and whether it will lift off the tops by the afternoon, say (diagram 25B).

Hill fog colder than $0°C$ can cause a growth of ice to build up on the windward sides of rocks and plants. This ice is known as *rime;* it is soft, white and feathery, and over several days can become many centimetres thick. It can harden by partly thawing and then refreezing. It differs from hard, glassy *black ice*, or glaze, formed by the freezing of rain or seepage.

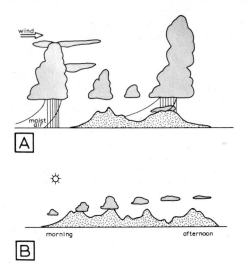

25. A. Passing patches of moist air formed in showers give short-lived clouds after lifting

    B. Cumulus cloud base often rises during the day

26. Clear Top

Even with a forecast of widespread haze, coming from faraway places, the tops can stay in clean air so long as the haze is trapped below an inversion aloft. This is most likely to happen in a winter high, where the air above the inversion has been warmed by slow but long-lived sinking. Such air can be so dry there is almost no moisture in it. With long hours of sunshine, and perhaps a wind as well, there is a risk of cracked and sun-burnt skin.

The haze top all around gives a false horizon, and you may be unable to see the sea, say, which you know to be only a few kilometres away, whereas high tops tens of kilometres away may stand out clearly.

Watch for a deepening of the haze layer, for it may swell up to the highest tops. Moreover, when that happens a cloud sheet may form near the haze top, or spread in from upwind, and the highest ground becomes enshrouded in cloud, so bringing an abrupt end to a gloriously sunny day on the tops (diagram 26).

haze layer

26.   Cloud sheet forms as haze layer deepens when crossing high ground

27. Rainy Top

Mountains are undoubtedly more rainy than low ground. This is not just because rains are more common there; they also tend to be heavier. Even so, most mountain clouds still give no rain.

A forecast of 'occasional rain or drizzle' for low ground can be a soaker in the hills. Most heavy falls over British mountains come with strong south-west winds ahead of a cold front (map 27A). The winds are warm and moist from low latitudes and they are gently rising to give widespread sheets of stratus or stratocumulus clouds say 1500m deep — enough to give outbreaks of light rain or drizzle. As the winds rise over the mountains, the clouds deepen, thicken and darken (diagram 27B). If there is not enough time for new raindrops to form (something like an hour is needed) before the wind has crossed the mountains, there will be little rain. But if there are rain and drizzle drops already in the clouds, as they come up to the windward slopes, they grow quickly in the wetter hill cloud to give a drenching rain that goes on and on until the cold front passes and the south-west winds are replaced by drier west or north-west winds. Some of the drops blow over to the leeward side if the highlands are not too broad. Such a drenching rain comes on steadily as the clouds deepen ahead of the front, both by lowering base and rising top, and may last five or ten hours if the cold front is slow-moving. The fall can be 50 to 100mm, or more, and there may be floods. Such falls can bring five or ten times as much rain as on low ground to windward. Maps 4B and 5B show examples of moist air streaming for hours over mountains to give heavy falls;

Rain drops falling from higher clouds near fronts and lows also grow in the thickened hill cloud, leading to heavier rain than over low ground (diagram 27C).

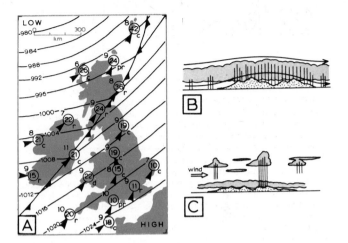

27. A. Typical weather map with strong, moist south-west winds giving heavy mountain rain

   B. Clouds already on the wind deepen, thicken and darken over the hills to give more rain than upwind

   C. Rain falling from higher, unseen clouds, and growing in thickened hill cloud beloww

28. Showery Top

A day with forecast 'light and scattered showers' can be a wet one in the hills because the showers turn out to be heavy and frequent. We have seen (Section 12) that showers fall from deep cumulus clouds, or the cumulonimbus that grow out of them, and that such clouds tend to be deeper and more frequent over mountains than over low ground (Section 22). It is no wonder, then, that mountain showers can be heavier and more frequent than over low ground (diagram 28A). Indeed, on days when showers are only just able to grow inside the clouds they are more likely to start over the mountains than over low ground. (Satellite photograph 4B shows an example of shower clouds starting to grow over the Alps and Pyrenees.) **Frequent light showers, each not amounting to much, can plague the** hills when low ground stays dry (diagram 28B). What is more, the showeriness is likely to be greater over broad highlands rather than narrow ridges because the wind will often carry shower clouds well away from isolated peaks during the time it takes for rain drops to grow (diagram 28C).

Expect greater showeriness in highland areas, the more so on their windward sides, when showers are forecast to be widespread; and expect showers to go on longer there when they have died away over low ground.

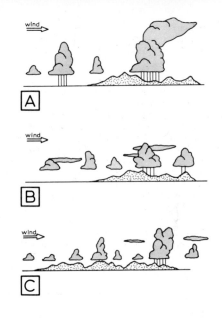

28. A. Mountain showers can be heavier and more frequent than over low ground

    B. Light showers can fall over the hills when low ground is dry

    C. Showers need not be heaviest on the windward side of highlands because rain drops take time to grow as they drift on the wind

29. Dry Top

It is rare for the tops to be dry when widespread rain or drizzle is forecast. It can happen if the mountains are tall enough to poke through the sheet of rain cloud. But we have seen that clouds need to be at least about 1000m deep to give even light drizzle, so British mountains are unlikely to poke through most rain clouds (diagram 29A). Such weather is difficult to forecast because the cloud top will be close to the summits of the highest mountains. Moreover, the top is likely to have vast, gentle waves a few hundred metres high that move by, enshrouding the summits from time to time (diagram 29B). Warm sectors are perhaps the most likely places to look for this kind of weather.

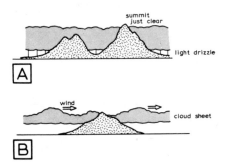

29. **A.** British mountain tops are unlikely to poke through clouds deep enough to give rain (or even drizzle)

   B. Passing waves in a sea of clouds cause summits to go in and out of cloud

30. Thundery Top

British mountains are not much more thundery than low ground. Any difference may be due more to distance from the coast than height above sea level. Even so, we should expect more thunderstorms over broad highlands compared with low ground because, as we have seen, they grow from vigorous showers, and showers tend to be heavier and more frequent over mountains. This is likely to be most true on summer days when storms are first appearing after a warm spell, say ahead of a cold front coming slowly eastwards from the Atlantic (map 7C), or on the edge of a low spreading north from France (map 14C). Torrential rain, and even hail, in these storms can give heavy flooding, and any washing away of a hill side is said to be due to a 'cloud burst'.

31. Snowy Top

When rain is forecast, the more so in winter, we need to judge the chances of *snow on high ground*. Much of the rain that falls near sea level at fronts and in lows starts high in the air as snow. Flakes may start 10,000m or more above sea level, take an hour or two to fall, and melt to rain drops only after reaching less than 1000 or 2000m above sea level. Flakes take a few minutes to melt as they fall into warmer air, so the *melting layer* is often up to 300m deep. In this layer the partly melted flakes are called *sleet*. When there is widespread rain inside cloud, the bottom of the melting layer has a temperature about $2°C$, for sleet in cloudy air is seldom seen at higher temperatures. If the temperature in a valley at 200m above sea level where it is raining is $4°C$, and up to the bottom of the melting layer there is the usual lapse rate of $1°C$ for every 200m, sleet or wet snow can be expected above about 600m, and dry snow above about 1000m (diagram 31). The forecaster can advise on where the melting layer will be.

Snow reaching the ground ahead of a winter front or low often changes to rain after a few hours as warmer air spreads in (see maps 3D and 15C). Sometimes, however, rain can change to snow. With long-lasting rain in light winds, the melting layer can deepen as the snow and sleet cool the air. In this way the snow can reach nearer and nearer sea level. A sign that this might be happening is a slow but steady fall of temperature towards $0°C$ as the rain goes on. In time a few flecks of sleet can be seen, then more and more until there are only snow flakes, wet because they have melted a little. The bottom of a melting layer has been known to fall 500m in this way over several hours.

If cloud base is below the melting layer, settled snow on the higher slopes cannot be seen from the valleys, so there is no obvious sign of the bad weather on high ground. But up there, with snow falling in cloud on to snow-covered ground, there can be a *white-out*, where shadows disappear so that avoiding snow-filled hollows become very difficult, and horizon judging is very hazardous, especially near the tops of crags.

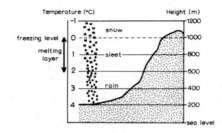

31.   A melting layer between 600 and 1,000m – snow above, rain below

32. Windy Valley

Winds in valleys can play strange tricks. The general forecast may speak of moderate speeds yet there may be strong or gale force winds in one valley but not in another nearby. There are many reasons why valley winds can be stronger than over open country. Here are some of them.

1. When the wind can hardly rise over a ridge it tends to blow strongly through any gap there may be (diagram 32A). Some mountain passes are named after their strong winds. On the leeward side the wind can shoot out as a strong jet like water through a sluice, or fan out into a moderate wind. A wind blowing around a hill is strengthened on the shoulders, sometimes to as much as twice the speed over open country (diagram 32B). This kind of strengthening is most likely when there is an inversion aloft lying below summit height, for there may not be enough energy in the wind to lift the inversion above summit height.

2. When the wind blows into a narrowing valley, there may be a strengthening up to, and more so over, the pass at the end of the valley (diagram 32C). Such a *funnelling* is common in deep valleys lying more or less along the wind. For example, a valley lying E-W is likely to have a west wind if the weather map shows a wind from between SW and NW, but an east wind if if the map shows NE to SE.

3. On sunny days when little or no wind is forecast, a gentle breeze (up to force 2) can start to blow up-valley during the morning, dying away towards or soon after sunset. Such *up-valley winds* can bring some relief on an otherwise hot, still day. They blow when there are *up-slope* (or *anabatic) winds* on sun-heated slopes, for the up-valley wind feeds them by drawing in air from open country (diagram 32D). At night there may be a *down-valley wind* (again up to force 2). It is fed by cool air from the slopes sliding down hill, either to collect in the valley bottom or to flow gently out if the valley lets it. Down-valley winds tend to come in pushes every 10 or 20 minutes (or even an hour or more) with calms between. Where valleys meet, the colder wind undercuts the less cold. Because both up-and down-valley winds are weak they are easily swamped by only moderate winds you might have expected from the weather map.

4. Sometimes surprisingly strong winds blow *across* valley. This happens when air flows over a ridge like water over a weir. The strengthening wind down the lee slope is known in the Eden valley, Cumbria, as a *helm wind*— a term that may be used

more widely for this kind of wind. Watch out for such a wind whenever an inversion aloft lies just above the ridge line (diagram 32E), the more so at night rather than by day, and in dull weather rather than sunny. Speeds can reach gale force although only moderate to fresh winds are forecast, but forecaster will be able to advise on whether a helm wind is likely. Evenso, remember a ridge seldom has a straight top; some parts may be too high for a helm wind, and moreover the inversion is likely to rise or fall. Don't wonder at a sudden cross-valley wind springing up when you thought, after looking at the weather map, you would be in a sheltered place. What's more, a cross-valley wind can bring small but strong whirls that form in the wakes of knolls and ridges. Over lakes these whirls can cause sudden and even alarming stirrings of the water surface.

It is very difficult to forecast just where and when these strong valley winds will be blowing. Expect them whenever the general forecast speaks of winds of more than moderate strength. Use your topographic map and the forecast weather map to judge where strong valley winds are likely to blow.

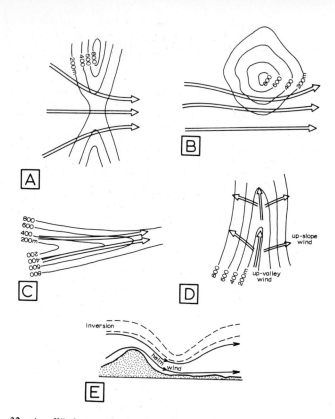

32. A. Wind strengthening through a gap in a ridge, causing a jet on the lee side

    B. Wind strengthening as it blows around the side of a hill

    C. Funnelling of wind along a narrowing valley

    D. Sun-heated slopes cause up-slope (anabatic) winds fed by an up-valley wind

    E. Strong lee-slope wind (helm wind) when a temperature inversion aloft lies not far above ridge height

33. Calm Valley

On days when strong winds are forecast, some valleys can still be calm. The most likely ones are those lying across wind, but as we have seen (Section 32), even there the wind can be remarkably strong. On many days when there is at least some sun, the wind blowing across a ridge leaves the ground near the ridge line, and the leeward valley is filled with a slowly overturning eddy (diagram 33A). Valley bottom winds are then not only light, they blow against the wind expected in the forecast. They contrast strongly with the wind at ridge height, such as is shown up by the racing clouds. These eddies, with their light winds, can give way at times to a helm wind, so a valley that might be thought to be sheltered turns out to have surprisingly varied winds, some strong enough to be uncomfortable, if not hazardous.

Even when a more or less steady wind blows along a valley, there can be calm places, such as behind a knoll (diagram 33B), a spur on the valley side, or just beyond a bend (diagram 33C). In all these places there can be eddies; the first has a level axis, the others have sloping ones. But all these eddies are likely to come and go; at times the wind spills over the knoll or around the spur or shoulder.

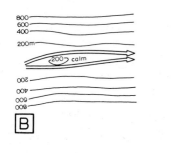

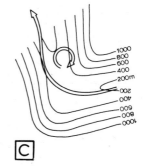

33. A.   Slowly-overturning eddy in a valley lying across wind
     B.   Calm patch in the lee of a valley-bottom knoll
     C.   Eddy just beyond a sharp bend in a valley

34. Cold Valley

A sheltered valley on a clear, calm night can be colder than open country by 5°C or even more. This is mainly because the wind over open country usually keeps the air stirred enough for cooling to be spread through a layer deeper than the valley. If the valley cools quickly as the wind drops to near calm around sunset, a very strong ground inversion can form. As cooling goes on through the night, some deep valleys can have a frost by sunrise, even in summer. When camping, therefore, it would be worthwhile seeking a place 100m above valley bottom so as to miss the coolest air at the bottom. Even a 50m difference can be 5°C warmer (see Section 20). On such clear, calm nights, sheltered valley bottoms are about as cool as each other no matter what their heights above sea level. Much the same thing can happen in sheltered hollows high in the hills; they are known as *frost hollows*.

In a steep-sided valley the sun may not be able to shine on the bottom, the more so in winter and when the valley lies east-west, not north-south. Such a valley can stay colder by day than open country. Moreover, snow lies longer and helps to keep the air cooler. If the cool air is unable to drain away easily, winds stay light and more or less cut off from winds at greater heights. The cold air in the valley bottom behaves something like water in a bath — it can surge backwards and forwards, or from side to side, so that on the hill slope cold air wells up and down more or less rhythmically, with periods of minutes or hours, depending in part on valley size (diagram 34A).

After a warm day with an up-valley wind, a down-valley wind can set in suddenly where cold air can begin to slide down the east-facing slopes even before sunset. At onset, as the wind direction reverses, the temperature can *fall* several degrees in as many minutes (diagram 34B). But at some places, where the down-valley wind starts late and after the daytime wind has dropped to calm, and a ground inversion formed, as the wind picks up there is a temperature rise.

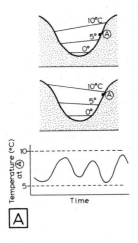

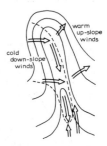

34. A. Valley-bottom temperature inversion surging to and fro,
        causing the temperature to rise and fall on the slopes

    B. Down-slope (katabatic) winds on the shaded, east-facing
        slope have started before up-slope winds on the opposite
        slope have stopped A cold, down-valley wind suddenly
        replaces a warm up-valley wind

35. Warm Valley

On some days a valley can be warmer than open country. There are four main ways this can happen.

1. On a sunny day, a valley floor becomes hotter than open country at the same height. This is because, although both places are heated by the sun at the same rate, the cooling by loss of heat into space is slower in the valley, for the walls make the sky there smaller than in open country (diagram 35A). Moreover, the walls radiate heat back to the floor. In a north-south valley, the west-facing wall is the warmer of the two at the time of highest temperature in the afternoon.

2. In a rain shadow, where the wind is blowing down from rainy hills, the air is warmer than at the same height on the windward side (diagram 35B). It became warmer as it lost some of its moisture as rain. The warm wind on the leeward side is called a *föhn wind*. It is warmest in valley bottoms. On the other hand, a wind crossing the hills but giving no rain is not warmed in this way.

3. When an inversion aloft lies below ridge height, winds below it may be unable to cross the ridge — they are blocked — so the wind sweeping down the leeward slope has come from above the inversion and it is warmed by compression as it sinks (diagram 35C). This warm wind is another kind of *föhn,* found not when the hills are rainy but even when the tops are cloudfree, above the inversion (Section 23).

4. If there is a föhn wind of either kind, the valley is likely to have broken cloud, so sunshine gives warmer weather than on the cloudy, windward side of the hills.

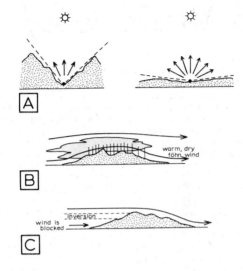

35. **A.** Rate of loss of radiant heat to space from the valley bottom is less than from open country; hence the valley becomes hotter by day

   B. A wind blowing down from rainy hills is warmer than on the windward side at the same height. It is one kind of föhn wind

   C. Warm wind blowing down a lee slopes from above a temperature inversion when the wind beneath is blocked. It is another kind of föhn wind

36. Dull Valley

Days when widespread broken clouds are forecast can be dull and sunless not only over the hills but also valleys running among them. By and large, we should think of the wind blowing over a highland area in one sweep and not much affected by valleys lying across the wind. Sheets of cloud formed by the lifting will then cover hill and dale alike. Slowly overturning eddies in cross valleys more or less stop the over-riding air from coming down into the valleys (diagram 36A). Cumulus clouds do much the same thing as they drift by in the wind, but because on a given day some hills are better than others at setting off or enlarging cumulus clouds, parts of a valley can stay sunless whereas other parts get fitful sun.

Sheets of stratus clouds over low country, capped by an inversion aloft, can be blocked as they are brought up to a ridge on the wind. Only where a valley lies into the wind can the low cloud spread among the hills. Indeed, such an exposed valley can stay dull and cool whereas another nearby one, lying across the wind, is sunny and warm with a föhn wind (diagram 36B). Cool, moist air feeding up a valley, after being heated over the hills, leads to cumulus clouds that are more likely on the windward slopes and have a lower base than around valleys not reached by the moist air.

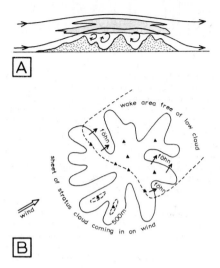

36. A. Valley eddies causing the wind to blow across highlands in more or less one sweep

    B. Highlands blocking an in-flowing sheet of stratus cloud, leading to sunny leeward valleys with föhn wind, but cool and dull windward valleys

37. Cloudy Valley

Apart from more or less broken cumulus and stratocumulus clouds, which we have seen are common over mountains (Section 22) and, by drifting on the wind, also over nearby valleys, there are other kinds of broken cloud found mostly over valleys.

1. When there is an eddy in a cross valley, and rain has been falling for hours so the valley air is very moist, then shreds of cloud can form in the upward-moving part of the eddy. These clouds are best seen where there is a dripping forest on the hillside, and very moist air is fitfully drawn out from among the trees (diagram 37A). Such shreds can be the first of much more widespread low clouds to form in a long spell of rain.

2. When a cold front passes overhead and the wind changes direction from along a valley to across it, then some low clouds from the warm air stream become trapped as the eddy forms. Such low clouds show up against the rising base of higher clouds behind the front (diagram 37B). These 'left-overs' go at last by mixing slowly with over-riding air, or because the air becomes warmer during the daytime.

3. When there is an inversion aloft, the first cumulus clouds to form during the morning can be over the lower slopes below the inversion, and the mountain tops stay clear (diagram 37C). Later, the sun's heat may destroy the inversion and cumulus clouds form over the tops as well.

4. When the wind blows across a ridge and sinks back on the leeward side it can overshoot and set up a train of waves, in the crests of which smooth, lens-shaped or banded clouds can form (diagram 37D and photograph 37E). These *wave clouds* (or lenticular clouds) are more or less fixed in the sky and the wind blows through them. Where there are many ridges with a great variety of heights and shapes, these clouds can be scattered over the sky, or in lines, or even piled one above the other. They can be a forerunner of cloudy or even wet weather, for they show there are sheets of moist air that may later be lifted to form clouds more widely as a front or low comes nearer. Beneath some waves the wind can turn right over as a *rotor* (diagram 37F). In the bottom of the rotor there can be strong gusts, blowing against the main wind stream, and they are especially dangerous on hill tops.

5. If there is already a sheet of cloud in a wind that is giving waves in the lee of a ridge, there can be gaps at the troughs of the waves. These *wave gaps,* too, are more or less fixed in the sky; hence places beneath them get much more sun than others only a kilometre away (diagram 37G and photograph

37H). Photograph 37I shows wave clouds and wave gaps seen by satellite in a sheet of stratocumulus cloud over the mountains of South Wales. On rare days, wave gaps in the lee of an isolated peak can be in the form of a 'V', like the wake of a moving ship. Such a wake is more easily seen from a satellite than from the ground; photograph 37J is an example from north-west Scotland.

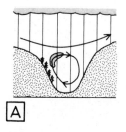

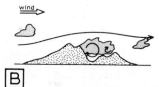

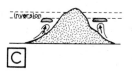

37. A. Moist eddy in a cross-wind valley causes cloud shreds to climb slowly up the windward side

   B. For a time after a cold front has passed, patches of low cloud from the earlier warm air may linger in the eddies

   C. The first morning cumulus clouds can be over the lower slopes when there is a temperature inversion over the upper slopes

   D. Wave clouds forming in the crests of air waves over and in the lee of a ridge

E. Lens-shaped and banded wave clouds over Snowdonia, base about 3,000m

F. Wave gap in the trough of a lee wave

G. Waves with a rotor, giving a patch of winds blowing against the main stream

H. Fitful sunshine on a day with two, thin, broken sheets of
    stratocumulus cloud and a wave gap. The wind blows
    down from cloud over the Glyders on the right, and rises
    into the dark wave cloud on the left

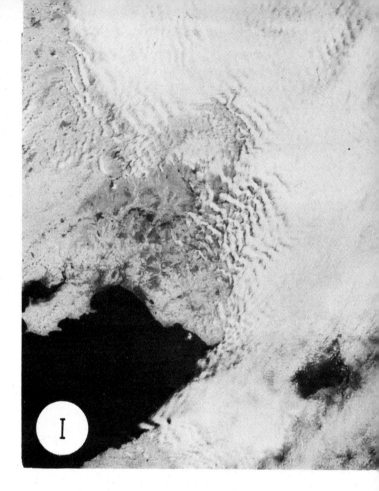

I. Satellite picture of cross-wind wave clouds in a shallow
sheet of stratocumulus clouds carried on north-east winds
over South Wales. The Gower peninsula, Swansea and
nearby areas are cloudfree, probably because the wind
blows down in the lee of the mountains. In contrast, the
cloud sheet passes less disturbed across the Severn estuary
to Exmoor

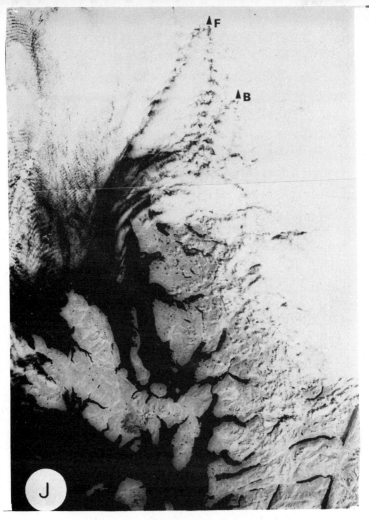

J.   Satellite picture of an extensive sheet of stratocumulus
     cloud with V-shaped wakes downwind of some mountains
     in north-west Scotland. The wakes consist of long chains
     of dark, cloudfree holes that would give fitful sunshine in
     otherwise dull weather. The clearest wakes are downwind
     of Foinaven (F) and Ben More Assynt (B) but others,
     perhaps caused by Suilven and nearby peaks, seem to be
     confused. Skye is cloudfree on the leeward side of the
     highlands

97

38. Cloudless Valley

A day forecast to be dull with an unbroken sheet of stratus cloud
can be sunny and more or less cloudless in a sheltered valley. This
happens when a ridge blocks the flow of clouds over the valley
because there is an inversion aloft. If cloud top is below ridge top,
no cloud may be seen from the leeward valley (diagram 38A). But if
clouds rise just above ridge top, they may be seen spilling over like
waterfalls, endlessly evaporating in the same place. Such a cloud on
the ridge is known as a *föhn wall* (diagram 38B). The quilt-like
pattern on its upper side can be seen sliding into its downwind edge.
Where a ridge varies in height along its length, there may be spillage
over the lowest parts, but blocking elsewhere (photograph 38C).
Moreover, if the cloud top wells up or down so the spillage grows
and shrinks.

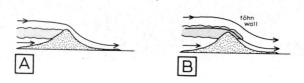

38. A.   Cloud sheet wholly blocked by a ridge

    B.   Cloud sheet partly blocked – the top pours over and
evaporates by compression at the föhn wall

C.   Stratus cloud pouring through the narrow head of the Llanberis Pass, illustrating the great contrast in weather there can be between nearby valleys when a cloud sheet is blocked

39. Hazy Valley

A day when clean air sweeps across country, from far away with few sources of smoke, can still be hazy in the valley if it is sheltered and there are smoke sources within it. Smoke may come from factories, or the house and rubbish fires of a town. With an inversion *aloft,* the smoke may be spread up to its base, perhaps 1000 or 2000m above the ground. But with a *ground* inversion the smoke may be taken up no more than 100m, to gather in blue or grey sheets as each plume first rises and then spreads sideways (diagram 39A). After a clear, calm night in winter, the lighting of morning fires can thicken the haze to a pall through which things can be seen up to only a kilometre or two away.

Such a haze can be cleared away if the wind changes direction to blow along the valley. It can be thinned by mixing over a greater depth, as can happen during the daytime, when the sun's heat sets up-slope winds blowing (diagram 39B) and cumulus clouds may form later.

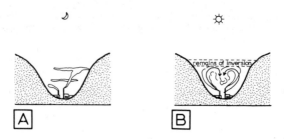

39. A.   Smoke plumes spreading within a valley-bottom temperature inversion

   B.   Smoke mixed upwards as the sun's heat destroys the temperature inversion

40. Foggy Valley

Fog is fickle. After a clear night with a forecast of fog, valleys can often be fog-free, or else there are only patches. Moisture in the air has condensed out as dew instead. The reason for this is probably that valley winds during the night do not fall light enough to let fog form. In open country, the wind needs to be no stronger than about Force 1 for fog to form, whereas down-valley winds on clear nights are often a little more than that, as can be seen in the drift of chimney smoke. Where the down-valley wind drops out to calm, say near barriers such as tree clumps or a narrowing of the valley, fog patches are likely. Where a stronger breeze blows in from a side valley, fog is unlikely.

Valleys are more likely to have fog when it can spread in from nearby sea or low ground. It might be thought that this was impossible because the wind ought to mix cold, foggy air with warm, dry air aloft, so that the fog would evaporate. But once widespread fog has formed, the ground inversion moves steadily to the fog top, and mixing with clear air above becomes more difficult. Even a Force 2 breeze can then carry fog. Any lifting needed to take the fog up into a highland valley may well add a little to the cooling of the foggy air. Watch the weather map for days when light winds blow from places with widespread fog.

Fog will not form over land if there is much cloud. Because some valleys can be more or less cloudless on a generally cloudy day (Sections 37 and 38), watch out for unforecast fog patches in those valleys.

41. Clear Valley

One valley can be fog-free when others nearby are fog-filled.
Because valley fog often spreads from elsewhere, a valley can be free
if it is sheltered by high ground from the fog-bearing breeze. This
can happen in much the same way as when a ridge blocks a stratus
cloud sheet (Section 38). Photograph 41A is a satellite picture of
shallow fog on north-west winds being blocked by the mountains of
south-west Scotland. If there is a föhn wind, blowing down from
above fog on the windward side of the ridge, valley fog is even less
likely (diagram 41B).

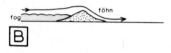

B. Blocked fog sheet and a lee-side föhn wind

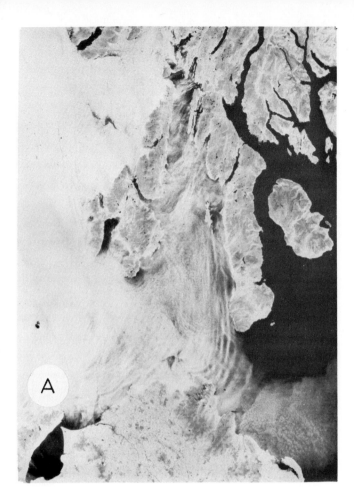

41. A. Satellite picture of extensive shallow sea fog pouring through the North Channel between the Mull of Kintyre and Northern Ireland. Fog fills the Firth of Lorne but the north-west wind is blocked by the mountainous islands of Islay and Jura. Fog also seems to be forming in the Sound of Jura as warm, moist air spreads across cool sea. Arran and the country around Loch Fyne are clearly seen through the cloudfree skies there

42. Rainy Valley

A rainy day in the hills is likely to be a rainy one in the valleys as well, because rain drops drift on the wind. Most drops fall at 10 to 20 kph. Because they have to fall at least a kilometre or two through and below the clouds where they started, they take some fraction of an hour to fall out, and in so doing will drift several kilometres on most winds. Most British valleys among the hills are no wider than this, so even if the rain cloud sits over the hills and does not move with the wind, the valley will still get rain, often plenty, but more in the form of small drops that can drift furthest (diagram 42A). This is perhaps most likely in the driving rain of a warm sector. When a low, fitful sun shines on the valley, a rainbow comes and goes. Snow flakes are carried even more easily on the wind because they fall slower — only a few kilometres an hour. Hence the rain that comes from their melting falls even further downwind.

Because some mountains are better able than others to form rain clouds, or to renew clouds already on the wind, it follows that clouds just in their lee will be wetter than over other valleys. For example, on a day with seemingly endless rain in a strong, moist south-west wind, there can often be seen, looking along a cross-wind valley, curtains of rain sweeping down on some parts more often and more heavily than on others. Likewise, valley heads are more often rainy than valley feet (diagram 42B).

Sometimes there are other, unseen rain or snow clouds at greater heights caused by winds rising over hills. Rain from these clouds can spread far down wind, even at 50 to 100 kph in the winds at heights of 3000 to 6000m (diagram 42C).

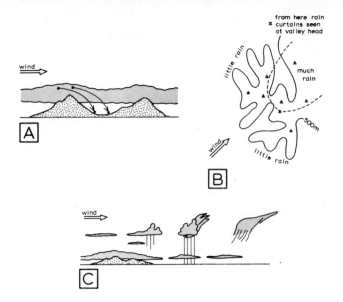

42. A. Curved paths of rain drops as they grow whilst drifting into a valley after starting well upwind

    B. Heavier rain over the central highlands can be seen up the valleys

    C. Upper clouds, set off by the mountains, can give rain far downwind

43. Showery Valley

When showers are forecast, valleys are likely to be as wet as the hills nearby. On the one hand, showers carried on the wind to a highland area can be heavier over the hills than in the valleys because the shower clouds are thickened as the wind rises over the hills. On the other hand, shower clouds that start over the hills are not likely to rain out much until some kilometres downwind, simply because the drops need a growth time during which the clouds drift from the hills. Sometimes the shower cloud itself may not start to form until downwind of a mountain. This can happen on a showery day when the wind splits and blows around both sides of the mountain, and then meets again along a line where the air is forced to rise and so cloud forms (diagram 43A). Photograph 6E shows a line of cumulus clouds, downwind of the Carneddau in North Wales, which may well have formed in this way.

Where the wind takes at least an hour or two to cross a broad highland, showers can lose all their drops before reaching the leeward side. This is why, for example, eastern Scotland and the Welsh Borders have little rain with a showery west wind (diagram 43B). Map 3A shows sheltering around Glasgow with north winds, and in map 15A there is sheltering over the Hebrides with east winds.

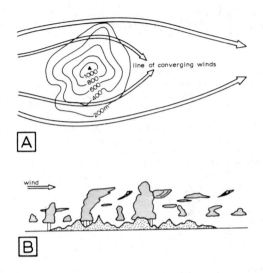

43. A. Lee-side line of converging winds where shower clouds are likely to grow

    B. Showers over broad highlands can die away before reaching the lee side

44. Dry Valley

On a day with rain or drizzle from long-lived stratus or stratocumulus clouds over broad highlands, the leeward side, more so the valleys, can have little or no rain. This happens when the wind is too weak to carry drops from the windward side where they form (diagram 44A). There can be falls of 10 or 20mm in the central highlands but almost nothing only ten kilometres away downwind. The dark, gloomy mountains then contrast strongly against the dry leeward valleys with their fitful sunshine. Looking from the valley towards the hills, not only can you see curtains of rain sweeping along in the wind, but also sometimes the downwind edges of the cloud mass are seen to be breaking endlessly over more or less the same places. Moreover, the base is higher than on the windward side. If the clouds are deep, only the lowest sheets may break, and some rain may go on falling from higher sheets (diagram 44B).

The dry leeward side is known as a *rain shadow*. It may not be quite dry, for some drops may come across from time to time (Section 42). Map 4B shows broken cloud and dry weather around Aberdeen when moist, cloudy west winds are giving long-lasting rain and drizzle over western Scotland.

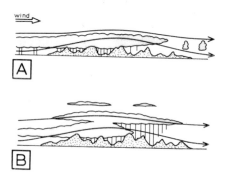

44. A. Weak winds may not be able to carry rain drops as far as the leeward side of broad highlands

B. There is a higher base on the leeward side but rain still falls from the upper cloud sheet

45. Thundery Valley

Just as showers can drift from hills to valleys, so can thunderstorms, because most are really no more than large and heavy showers. On days when thunder is forecast, don't expect the valleys to escape. However, some valleys, or parts of valleys, may get more thunder than others if storms are more likely to grow over or near some mountains than others. Storms are sometimes said to 'move in circles,' but this would be most unusual. Two or more short-lived storms moving along different paths might give this impression.

46. Snowy Valley

Falling snow in a valley is unusual when only rain is forecast for open country at the same height. It can happen when cold air has been trapped in a sheltered valley, say ahead of a warm front in winter when winds are light. Such snow is likely to be short-lived.

REMEMBER

- Mountain weather is fickle, and its changes are difficult to forecast.

- Weather on the summits can be unexpectedly severe, particularly on the higher and more northerly summits of Scotland.

- Before planning a day among the mountains get a forecast and judge if the weather is likely to be favourable.

- Before starting out, get the latest and most detailed forecast. If one is not available on the automatic telephone weather service, get it from the local forecast office whose number is given in the telephone directory. Change your plan if the weather is likely to turn foul.

- Whilst outdoors, take an interest in the weather, and watch out for any marked differences from the forecast. Change your plan if the weather is turning foul.

Be weather wise!

FURTHER READING

INTRODUCTIONS

Know the Weather C. E. Wallington and D. E. Pedgley
   E P Publishing Ltd (1979)

The Weather Sir G. Sutton
   Teach Yourself Books (1974)

The Weather Guide A. G. Forsdyke
   Hamlyn (1969)

MORE ADVANCED BOOKS

Atmosphere, weather and climate R. G. Barry and R. J. Chorley
   Methuen (3rd edn 1976)

Clouds and Weather R. K. Pilsbury
   Batsford (1969)

Cloud Study F. H. Ludlam and R. S. Scorer
   Murray (1957)

The Climate of the British Isles (eds) T. J. Chandler and S. Gregory
   Longmans (1976)